HSP *Georgia*
Science

CRCT Practice Tests
Grade 1
Teacher Edition

Harcourt
SCHOOL PUBLISHERS

Visit *The Learning Site!*
www.harcourtschool.com

Contents

Overview of *CRCT Practice*

<u>Practice Tests:</u> Two practice tests are provided for each unit (Earth Science, Physical Science, and Life Science) in *Georgia HSP Science* by Harcourt School Publishers. Each practice test consists of 10 items that address the Georgia Performance Standards covered in the corresponding units of *Georgia HSP Science*. These two tests for each unit could also be used as unit pretests and posttests or as a prescription for students who show a weakness in one of the domains of science. In addition, there is a 6-question unit practice test at the end of each unit in the *Georgia HSP Science* Student Edition.

<u>Cumulative Practice Tests:</u> Two cumulative practice tests are provided in this resource. Each practice test consists of 30 items that address all the GPS content standards for Earth, Physical, and Life Science, as outlined in the *CRCT Content Descriptions* publication from the Georgia Department of Education. In addition, the content, skills, and abilities associated with the Characteristics of Science are integrated into the three content domains. Each test is divided into two 15-question sessions. If you plan to give both sessions of a test on the same day, allow for a break between the sessions. These tests can be used at the beginning and end of the year to measure student growth. They also could be used to acquaint students with the format and style of the Georgia Science CRCT and to teach test-taking strategies. In addition, a 20-question cumulative practice test can be found on pages 252–257 in the *Georgia HSP Science* Student Edition.

Overview of the Georgia Science CRCT

The Georgia Science Criterion-Referenced Content Test (CRCT) is an achievement test. Students are assessed to determine their mastery of the content and skills set forth in the Georgia Performance Standards for science. Currently, the science test is given in spring to students in grades 3–8.

What content is assessed on the Georgia Science CRCT?	What kinds of questions are used on the Georgia Science CRCT?
The Georgia Performance Standards are grouped into three domains. **A. Earth Science** **B. Physical Science** **C. Life Science** The content, skills, and abilities associated with the Characteristics of Science are integrated into the three content domains.	**Multiple-Choice Items** Currently, the test includes only multiple-choice items. The number of items answered correctly is converted to a scale score.

The following two pages also appear in the *CRCT Practice* Student Editions. These pages familiarize students with the types of questions found on the Georgia Science CRCT and offer information to help students improve their performance.

About Tests

Taking a test is one way to show what you know. Each time you take a test, remember to:

- Listen to the teacher's directions.
- Read carefully.
- Mark or write the answers clearly.
- Work carefully.
- Do the best you can.

A sample question is below.

---

6. Look at the picture. How can you describe this animal?

A. It can move from water to land.

B. It has scales and gills.

C. It has wings and feathers.

`S1L1d`

Your answer should look like this.

6. Ⓐ Ⓑ Ⓒ

Test-Taking Tips

☑ **Use this checklist to help you prepare for the CRCT.**

☐ Do not stay up late the night before the test.

☐ Eat a good breakfast before the test.

☐ Use the restroom before the test begins.

☐ Bring more than one pencil.

☐ Know how much time you have to take the test.

☐ Stay relaxed during the test.

☐ Do not talk to other students.

☐ Raise your hand if you have questions.

☐ Do not get worried during the test.

☐ Pace yourself. Work quickly but carefully.

☐ Make sure you put your answer in the correct place.

☐ Do your best on the test.

Student Name:_____

Teacher:_____

Earth Science

1 (A) (B) (C)	5 (A) (B) (C)	9 (A) (B) (C)
2 (A) (B) (C)	6 (A) (B) (C)	10 (A) (B) (C)
3 (A) (B) (C)	7 (A) (B) (C)	
4 (A) (B) (C)	8 (A) (B) (C)	

Physical Science

1 (A) (B) (C)	5 (A) (B) (C)	9 (A) (B) (C)
2 (A) (B) (C)	6 (A) (B) (C)	10 (A) (B) (C)
3 (A) (B) (C)	7 (A) (B) (C)	
4 (A) (B) (C)	8 (A) (B) (C)	

Life Science

1 (A) (B) (C)	5 (A) (B) (C)	9 (A) (B) (C)
2 (A) (B) (C)	6 (A) (B) (C)	10 (A) (B) (C)
3 (A) (B) (C)	7 (A) (B) (C)	
4 (A) (B) (C)	8 (A) (B) (C)	

Student Name:_____

Teacher:_____

Cumulative Test

Session 1

1 (A) (B) (C)	5 (A) (B) (C)	9 (A) (B) (C)	13 (A) (B) (C)
2 (A) (B) (C)	6 (A) (B) (C)	10 (A) (B) (C)	14 (A) (B) (C)
3 (A) (B) (C)	7 (A) (B) (C)	11 (A) (B) (C)	15 (A) (B) (C)
4 (A) (B) (C)	8 (A) (B) (C)	12 (A) (B) (C)	

Session 2

16 (A) (B) (C)	20 (A) (B) (C)	24 (A) (B) (C)	28 (A) (B) (C)
17 (A) (B) (C)	21 (A) (B) (C)	25 (A) (B) (C)	29 (A) (B) (C)
18 (A) (B) (C)	22 (A) (B) (C)	26 (A) (B) (C)	30 (A) (B) (C)
19 (A) (B) (C)	23 (A) (B) (C)	27 (A) (B) (C)	

Name _____ Date _____

Earth Science Practice Test: Form A

1. Look at the picture. What is happening to the ice cube?

S1E2a

- A. It is melting.
- B. It is evaporating.
- C. It is freezing.

2. Which is the hottest season of the year?

S1E1c

- A. spring
- B. fall
- C. summer

3. It is winter. Snow is falling. Which word best describes the weather?

S1E1a, S1E1c

- A. cold
- B. warm
- C. rainy

Name _____ Date _____

4. Which of these is liquid precipitation?

S1E2b

- A. ice
- B. rain
- C. snow

5. Which of these tools shows the direction of the wind?

S1E1b, S1CS3a

A.   B.   C.

6. It is a warm, sunny day. What will happen to the water in an open container?

S1E2d

- A. It will condense.
- B. It will evaporate.
- C. It will melt.

7. Which of these is water in a solid form?

S1E2b

- A. rain
- B. a puddle
- C. hail

Name _____   Date _____

8. Look at the picture. The ice in glass A weighs 10 ounces. The ice melts to make the water in glass B. How much does the water weigh?

A. 6 ounces

B. 8 ounces

C. 10 ounces

S1E2c, S1CS4b, S1CS5a

9. Which of these tools measures temperature?

A. thermometer

B. rain gauge

C. balance

S1E1b, S1CS3a, S1CS7c

10. It is a winter day. It is very cold outside. What will happen to the water in a puddle?

A. It will condense.

B. It will evaporate.

C. It will freeze.

S1E2d

Name _____ Date _____

Earth Science Practice Test: Form B

1. Which of these is solid precipitation?

S1E2b

(A.) snow

B. water

C. rain

2. You have some water. The water freezes and turns into ice. How much will the ice weigh?

S2E2c, S1CS4b, S1CS9a

A. It will weigh more than the water.

(B.) It will weigh the same as the water.

C. It will weigh less than the water.

3. Which of these tools measures rainfall?

S1E1b, S1CS3a

A.

(B.)

C.

CRCT 5

Name _____ Date _____

4. It was cold for three months. Now it is warmer outside. What season is it?

S1E1c

A. summer

B. fall

(C.) spring

5. It is summer. The sun is shining. Which word best describes the weather?

S1E1a

A. cold

B. snowy

(C.) hot

6. Look at the picture. Rain has fallen on the house and yard. It is a warm summer day. What will happen to the water in the puddles?

S1E2d

A. It will condense.

(B.) It will evaporate.

C. It will melt.

CRCT 6

CRCT 5–6

Name _____ Date _____

7. Which of these does NOT tell about the weather?

A. cloudy

B. rainy

C. happy

S1E1a

8. A lake is frozen. It gets warmer outside. What will happen to the lake?

A. The ice on the lake will melt.

B. More lake water will freeze.

C. Waves will form in the lake.

S1E2a

9. In the water cycle, what happens AFTER the sun warms the water?

A. The water flows.

B. It rains.

C. The water evaporates.

S1E2d

10. Which thermometer shows a temperature of 80°F?

A.   B.   C.

S1E1b, S1C53a, S1C57c

Name _____ Date _____

Physical Science Practice Test: Form A

1. Which of these objects gives off light?

S1P1a

(A) a flashlight
B. an oven
C. a heater

2. Look at the picture. What causes the strings of a guitar to make sound?

S1P1c, S1C54a

A. volume
B. pitch
(C) vibration

3. You put together two different magnet poles. What will happen?

S1P2a

A. They will break.
B. They will repel each other.
(C) They will attract each other.

4. A wind chime makes a sound. What kind of pitch does the sound have?

S1P1d

(A) high
B. quiet
C. low

Name _____ Date _____

5. Amy is holding these objects. Which object will a magnet attract?

S1P2b

A. a plastic cup
B. an eraser
(C) a steel paper clip

6. Which of these objects will block light and make a shadow?

S1P1b

A. a glass window
(B) a wooden door
C. a piece of clear plastic

7. Look at the picture. Which sentence is TRUE?

S1P2a, S1C54a

A. Wood blocks magnetic force.
(B) Your hand does not block magnetic force.
C. Air blocks magnetic force.

Name _____ Date _____

8. Which of these sounds helps keep us safe?

Ⓐ a fire alarm

B. the wind blowing

C. a bird chirping

`S1P1e`

9. What is making the shadows in this picture?

A. The sun is too bright.

B. The sun is not giving off any light.

Ⓒ The objects are blocking the sun's light.

`S1P1b`

10. What makes sounds different?

Ⓐ pitch and volume

B. quiet and soft

C. high sounds

`S1P1d`

Physical Science • Form A

CRCT 11

Name _____ Date _____

Physical Science Practice Test: Form B

1. Look at the picture. What will happen when these two magnets are put together?

S1P2a, S1C54a

Ⓐ.They will repel.

B. They will attract.

C. They will give off light.

2. Max plays the violin. What happens when he rubs his bow across the strings?

S1P1c

A. They vibrate and make sound.

Ⓑ.They vibrate and the sound stops.

C. They break and make no sound.

3. Which sentence is NOT true?

S1P2c

A. Magnetic force can pull things through air.

B. Magnetic force can pull things through water.

Ⓒ.Magnets attract things no matter what they are.

Name _____ Date _____

4. Look at the pictures. Which makes the sound with the lowest pitch?

S1P1d

A. the bird

B. the bell

Ⓒ.the drum

5. Kim heard this sound. It kept her from walking into the street. What sound helped keep her safe?

S1P1e

A. a fan blowing

B. a cat purring

Ⓒ.a car horn honking

6. What is TRUE about the sound of a person yelling?

S1P1d

Ⓐ.It has a loud volume.

B. It has a soft volume.

C. Its vibrations will not make sound.

Name _____ Date _____

7. Which of these gives off light?

A. B. C.

S1P1a

8. Mike has these objects in his pocket. Which object is attracted to a magnet?

A. a crayon

B. a nail

C. a rubber ball

S1P2b

9. What is made when an object blocks light?

A. a shadow

B. magnetic force

C. sunlight

S1P1b

10. Which two magnets will attract each other?

A.

B.

C.

S1P2a, S1CS4a

Life Science Practice Test: Form A

1. Hunter saw a frog and a rabbit. How can he compare them?

S1L1d

(A.)They move the same.

B. They look the same.

C. They grow the same.

2. Look at the picture. Which part of the plant holds it up?

A flower

B leaves

C stem

A. part A

B. part B

(C.)part C

3. Why does a plant need light?

S1L1c

S1L1a

(A.)to make food

B. to make water

C. to get air

---

4. Which part of a plant makes seeds?

S1L1c

(A.)flower

B. leaf

C. roots

5. How are penguins, owls, and eagles alike?

S1L1d

A. They swim.

B. They breathe with gills.

(C.)They have feathers.

6. Look at the picture. How can you describe this animal?

S1L1d, S1C54a

A. It can move from water to land.

(B.)It has scales and gills.

C. It has wings and feathers.

7. Which body parts do animals use to get air?

S1L1d

A. feet or hands

B. fins or feathers

(C.)gills or lungs

Name _____ Date _____

8. Why do animals need food and water?

S1L1b

   A. to fly

   B. to get air

   C. to live and grow

9. Look at the picture. How does this animal move?

S1L1d, S1CS4a

   A. It swims.

   B. It flies.

   C. It crawls.

10. Jim wants to grow a tomato plant. What does his plant need?

S1L1a

   A. flowers, leaves, and fruit

   B. light, air, water, and nutrients

   C. roots and leaves

Name _____ Date _____

Life Science Practice Test: Form B

1. Why do plants need soil?

A. Soil has worms.

B. Soil can be hard or soft.

C. Soil holds water, which plants need to live.

`S1L1a`

2. How do fish use gills?

A. to swim

B. to get air

C. to find food

`S1L1b, S1L1d`

3. What part of a plant takes in light and air?

A. stem

B. roots

C. leaves

`S1L1c`

Name _____ Date _____

4. Look at the picture. How is this animal using the hole in the tree?

A. as shelter

B. as food

C. as air

`S1L1b, S1CS4a`

5. Look at the picture. What part of the plant does the arrow point to?

A. stem

B. seed

C. flower

`S1L1c`

6. Where do plants get the water they need to grow?

A. flowers

B. the air

C. the soil

`S1L1a`

Name _____ Date _____

7. Why do animals need air?

A. to grow and live

B. to be safe

C. to hide water

S1L1b

8. What do a plant's roots take in?

A. soil

B. food

C. water

S1L1c

9. How are a butterfly and a caterpillar alike?

A. They move the same.

B. They grow the same.

C. They look the same.

S1L1d, S1CS4a, S1CS5a

10. What do a plant's flowers do?

A. make roots

B. make seeds

C. hold it up

S1L1c

Name _____ Date _____

Cumulative Practice Test: Form A, Session 1

1. Find the arrow. What does this part help the plant do?

A. It takes in light.
B. It holds the plant in the soil.
C. It makes food for the plant.

2. These objects make sounds. Which sound helps keep us safe?

A. a clarinet
B. a smoke alarm
C. a bird

3. What does water become when it freezes?

A. rain
B. ice
C. clouds

Name _____ Date _____

and water

4. How do sunlight, air, nutrients, and water help plants?

A. They help plants move.
B. They help plants make water.
C. They help plants live and grow.

5. Compare the animals in the picture. How are they alike?

A. They both have gills.
B. They both have feathers.
C. They both have scales.

6. Which of these makes a soft sound?

A. a whisper
B. a bass drum
C. a siren

7. Look at the picture. Which object is attracted to the magnet?

   A. the sailboat

   B. the rock

   C. the paper clip

`S1P2b`

8. Sam put some water in a glass. The water was gone in a few days. What happened to the water?

   A. It melted.

   B. It condensed.

   C. It evaporated.

`S1E2d`

9. Which is another name for melted ice?

   A. snow

   B. water vapor

   C. water

`S1E2a`

---

10. Look at the picture. What kind of solid precipitation is falling outside the window?

   A. rain

   B. hail

   C. snow

`S1E2b`

11. You are growing a plant. Which does your plant need to make its food?

   A. seeds

   B. rocks

   C. sunlight

`S1L1a`

12. Which sound is soft or quiet?

   A. a siren

   B. an airplane

   C. a whisper

`S1P1d`

13. How would you describe a bear?

   A. It has fur.

   B. It has scaly skin.

   C. It has feathers.

`S1L1d`

Name _____ Date _____

14. Look at the picture. What do these thermometers show?

in shade   in sun

Ⓐ It is cooler in the shade than in the sun.

B. It is cooler in the sun than in the shade.

C. It is warmer in the shade than in the sun.

S1E1b, S1C53a, S1C57c

15. A magnet can NOT pull a toy through paper. What do you know?

A. The toy is a magnet.

B. The magnet is very strong.

Ⓒ The toy is not made of iron or steel.

S1P2c

Name _____ Date _____

Cumulative Practice Test: Form A, Session 2

16. Why does plucking a guitar string make sound?

A. The guitar is magnetic.

B. The string vibrates.

C. The guitar gives off light.

`S1P1c`

17. Look at the picture. Why is the girl in a shadow?

A. The sun shines through the windows.

B. The building makes the street warm.

C. The building blocks the sun's light.

`S1P1b`

18. What are sunny, cloudy, and windy?

A. kinds of rocks

B. kinds of weather

C. kinds of clouds

`S1E1a`

Name _____ Date _____

19. Rob wants to know which direction the wind is blowing from. What tool should he use?

A. a rain gauge

B. a wind vane

C. a thermometer

`S1E1b, S1C3a`

20. You see the shadow of a tree on the ground. What object must be blocking the light?

A. the sun

B. a fence

C. a tree

`S1P1a, S1P1b`

21. Look at these two magnets. What will happen if you try to put them together?

A. They will attract.

B. They will move fast.

C. They will repel.

`S1P2a, S1C54a`

Name _____ Date _____

22. What do all animals need to live and grow?

A. light, water, food, and air

B. light, nutrients, and shelter

C. air, water, food, and shelter

S1L1b

23. Look at the picture. Which letter shows an adult frog?

A. A

B. B

C. C

S1L1d

24. It is summer. The sun is shining. Which word best describes the weather?

A. cold

B. snowy

C. hot

S1E1a

Name _____ Date _____

25. It is a warm spring day. What will happen to the puddle in this picture?

A. It will freeze.

B. It will evaporate.

C. It will melt.

S1E2d

26. Which sentence is NOT true?

A. Magnetic force can pull things through air.

B. Magnets attract things no matter what they are.

C. Magnetic force can pass through your hand.

S1P2c

27. Summer was hot. Now it is getting cooler outside. What season is it?

A. summer

B. fall

C. winter

S1E1c

28. There are low dark clouds. The sun is not out. Liquid precipitation is falling from the sky. What word best describes the weather?

A. windy

B. sunny

C. rainy

S1E1a

29. Look at the picture. The ice in glass A weighs 8 ounces. The ice melts to make the water in glass B. How much does the water weigh?

A   B

A. 6 ounces

B. 8 ounces

C. 10 ounces

S1E2c, S1C5 4b, S1C5 5a

30. Which of these objects is a source of light?

A. a campfire

B. a guitar

C. a magnet

S1P1a

Name _____ Date _____

Cumulative Practice Test: Form B, Session 1

1. What do plants need from the soil?

A. air

B. food

C. water

`S1L1a`

2. Which form of precipitation is hail?

A. solid

B. liquid

C. water vapor

`S1E2b`

3. What is making the shadows in this picture?

A. The sun is too bright.

B. The sun is not giving off any light.

C. The objects are blocking the sun's light.

`S1P1b, S1CS4a`

Name _____ Date _____

4. What happens to water when it disappears into the air?

A. It melts.

B. It freezes.

C. It evaporates.

`S1E2d`

5. You have some ice. The ice melts and turns into water. How much will the water weigh?

A. It will weigh more than the ice.

B. It will weigh the same as the ice.

C. It will weigh less than the ice.

`S2E2c, S1CS4b, S1CS5a`

6. Look at the picture. You put these two magnets together. What happens?

A. They attract.

B. They repel.

C. They give off light.

`S1P2a, S1CS4a`

Name _____ Date _____

7. Look at the picture. How does this plant part help the plant?

S1L1c

A. It holds the plant up.
B. It takes in nutrients.
C. It takes in sunlight.

8. Which of these objects is attracted to a magnet?

S1P2b

A. a plastic ruler
B. a wood pencil
C. a steel paper clip

9. A lake is frozen. It gets warm outside. What will happen to the lake?

S1E2a

A. The ice on the lake will melt.
B. More lake water will freeze.
C. Waves will form in the lake.

© Harcourt

Name _____ Date _____

10. Which of these makes a loud sound?

S1P1d

A. a cat purring
B. a whisper
C. a siren

11. Look at the picture. Which object is a source of light?

S1P1a, S1CS4a

A. the lamp
B. the toy
C. the trophy

12. Lee saw a bird and a butterfly in his yard. How can he compare them?

S1L1d

A. They move the same.
B. They look the same.
C. They grow the same.

© Harcourt

Name _____   Date _____

13. Look at the picture. What kind of solid precipitation has fallen?

(A) snow

B. sleet

C. hail

`S1E2b`

14. Why does a plant need light?

(A) to make food

B. to make water

C. to get air

`S1L1a`

15. Keri beats her drum. What happens?

A. It vibrates and makes sound.

B. It vibrates and the sound stops.

(C) It gives off light.

`S1P1c`

Cumulative Practice Test: Form B, Session 2

16. These objects make sounds. Which sound helps keep us safe?

S1P1e

Ⓐ a bike bell

B. a stop sign

C. a car tire

17. How do fish use gills?

S1L1b

A. to swim

Ⓑ to get air

C. to find food

18. Look at the picture. What season is it?

S1E1c

A. fall

B. spring

Ⓑ summer

© Harcourt

---

19. Which tool can help you measure temperature?

S1E1b, S1CS3a

Ⓐ.

B.

C.

20. What part of a plant makes seeds?

S1L1c

A. the stem

B. the roots

Ⓑ the flower

21. Why do animals need air?

S1L1b

Ⓐ to grow and live

B. to be safe

C. to hide water

© Harcourt

Name _____ Date _____

22. What does ice become when it melts?

(A.) water

B. snow

C. hail

`S1E2a`

23. A magnet will pull a paper clip through water. What do you know?

A. A paper clip is not attracted to a magnet.

B. Water blocks magnetic force.

(C.) Water does not block magnetic force.

`S1P2c`

24. Look at the picture. What will this animal look like when it is an adult?

A. a frog

B. a fly

(C.) a butterfly

`S1L1d`

Name _____ Date _____

25. You hit a water glass with a spoon. Why do you make sound?

A. The spoon is magnetic.

B. The glass gives off light.

(C.) The glass vibrates.

`S1P1c`

26. What do animals need to live and grow?

A. food, water, sunlight, and air

(B.) food, water, air, and shelter

C. food, water, nutrients, and sunlight

`S1L1b`

27. What are rainy, cloudy, and snowy?

A. kinds of rocks

(B.) kinds of weather

C. kinds of clouds

`S1E1a`

28. How are gills and lungs the same?

A. They both help animals eat.

(B.) They both help animals get air.

C. They both help animals move.

`S1L1d`

Name _____ Date _____

29. Which tool shows how much rain has fallen?

A.

B.

C.

S1E1b

30. It is winter. Snow is falling. Which word best describes the weather?

A. cold

B. warm

C. rainy

S1E1a S1E1c